Orchids: How to Keep This Alive

HOW TO KEEP THIS ALIVE
ORCHIDS

Julianne Robertson

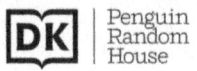

First published in the United Kingdom in 2026 by

August Books, an imprint of
Canelo Digital Publishing Limited,
20 Vauxhall Bridge Road,
London SW1V 2SA
United Kingdom

A Penguin Random House Company
The authorised representative in the EEA is Dorling Kindersley Verlag GmbH. Arnulfstr. 124, 80636 Munich, Germany

Copyright © Julianne Robertson 2026

The moral right of Julianne Robertson to be identified as the creator of this work has been asserted in accordance with the Copyright, Designs and Patents Act, 1988.
All rights reserved. No part of this publication may be reproduced or transmitted in any form or by any means, electronic or mechanical, including photocopy, recording, or any information storage and retrieval system, without permission in writing from the publisher.
No part of this book may be used or reproduced in any manner for the purpose of training artificial intelligence technologies or systems. In accordance with Article 4(3) of the DSM Directive 2019/790, Canelo expressly reserves this work from the text and data mining exception.

A CIP catalogue record for this book is available from the British Library.

ISBN 978 1 83598 253 2

Cover design by Ifan Bates

Cover images © RedGate Arts

Printed and bound in Great Britain by Clays Ltd, Elcograf S.p.A.

Look for more great books at
www.augustbooks.co | www.dk.com

Contents

INTRODUCTION ix
So you're the proud new owner of an orchid...

1. **WHAT IS AN ORCHID?** 1
   A plant's path from ancient medicine, to Orchidmania, to trending on TikTok

2. **HOW TO KILL THIS PLANT** 21
   Tips and tricks to ensure your orchid's certain death

3. **HOW TO KEEP THIS PLANT ALIVE** 49
   Re-blooming, repotting and becoming a modern-day orchid collector

4. **OTHER COMMON ORCHIDS** 67
   More orchids you can give to others (or yourself), and how not to kill them

CONCLUSION 75
Enjoy your plant, keep it simple and avoid botanical brutality

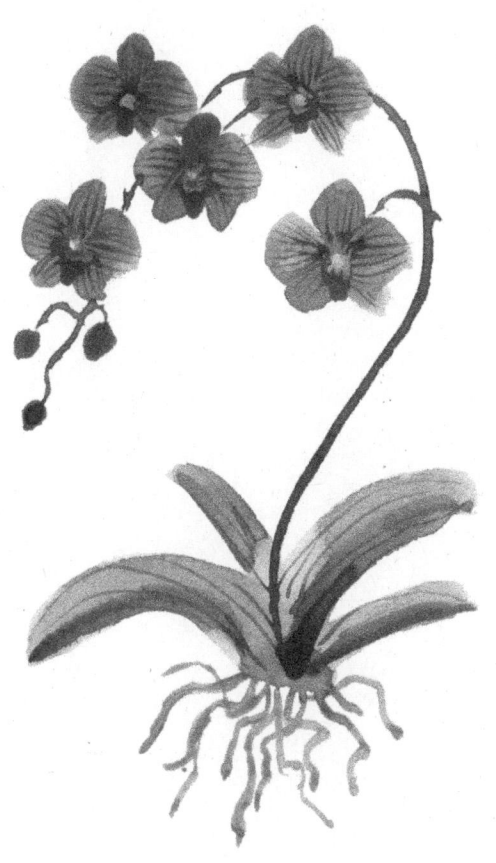

INTRODUCTION

So you're the proud new owner of an orchid…

Congratulations! You now possess one of the most rewarding and beautiful plants you can grow at home. Orchids are exotic, beautiful, varied and, with a little care and attention, they can be a long-lasting and rewarding addition to your home.

Whether you've been gifted this orchid, or you've picked it up as a treat for yourself, I know what you're asking yourself: **how do I keep this alive?**

That's where this book comes in: it's an easy-to-use guide to keeping your new plant healthy and happy. It will help you to understand the origins of orchids, the rich variety of orchids which exist and the conditions they need to thrive, including information on the right soil, light, water and location for your plant.

It will also take you through a few things **not** to do: the obvious no-nos and the honest mistakes which could harm your plant, or lead to its untimely demise. Alternatively, you can consider this a "how-to" guide, if in fact you're a pathological plant killer.

If you're a beginner to plant care, don't worry, because this book will also provide you with the basic knowledge you need to understand the language and practices of indoor gardening. It's not tricky or especially demanding – orchids in particular are low-maintenance plants – but it can be addictive; once you see your orchid growing, flowering and thriving, be prepared to want more!

1

WHAT IS AN ORCHID?

A plant's path from ancient medicine, to Orchidmania, to trending on TikTok

Phalaenopsis, also known as moth orchids, are among the most common plants to give as gifts, usually found in garden centres, supermarkets and even some high street shops, where they are sold in full flower. They're exotic, often brightly coloured… and perhaps a little intimidating? How can something so tropical be easy to look after in my home? Aren't houseplants tricky? Where are the instructions for this thing?

In fact, moth orchids are among the easiest of their kind to grow and aren't very demanding indoor plants. They can also be long-lasting and it's not difficult to encourage them to flower again and again.

The key to keeping this plant alive (or any plant, for that matter) is understanding it: where it comes from, what it likes and what not to do. This often varies between plants, sometimes even between varieties of the same species – what works for one may not work for another. It's always worth finding out a bit about your specific plant and its preferred growing conditions from the outset.

Let's take a closer look at the origins and typical growing conditions of Phalaenopsis, the moth orchid.

History of the Orchid

Orchids are currently one of the most popular houseplants, but they're not a modern phenomenon. The popularity of orchids stretches back hundreds, even thousands of years. There are records of orchids being used in Chinese medicine 4000 years ago to treat a range of complaints, including fever and infection.

The 19th century is when they became really desirable, and collectors were searching for orchids in earnest, travelling the globe to bring back new and interesting specimens to fill the glasshouses and conservatories of wealthy patrons. The obsession with this new exotic plant soon triggered a social craze known as Orchidmania: at its peak, orchids were such a status symbol that estate owners would buy and sell rare specimens for the price of a house.

Orchidmania takes hold!

In 1833, the Duke of Devonshire attended an exhibition at the Horticultural Society of London. He took one look at the butterfly orchid (*Oncidium papilio*) and decided to create the largest collection of orchids in the world.

Wealthy landowners followed suit, acquiring exotic orchids from around the world, and by the 1850s what had begun as one man's enthusiasm snowballed into Orchidmania.

Orchids soon became a status symbol, with orchid houses and conservatories on large estates displaying the latest acquisitions. This obsession wasn't cheap either – plants had to be collected from the wild as there was no method of mass production, so single specimens were changing hands for thousands of pounds. Hunter-collectors were travelling hundreds of miles to tropical countries to find new flowers and transport them back to their employers, at great risk to themselves and to the plants, which could easily be damaged or destroyed during the sea voyage home by disease, insects, rats or neglect.

Many of these vast collections were wiped out during the First World War; austerity and changing tastes in gardening during the following years meant a continued lull in the orchid trade.

Eventually, techniques were developed in the 1960s to mass produce orchids commercially, so the prices of plants went down and their accessibility and general popularity grew. Nowadays, the relative ease of sourcing, growing and transporting all kinds of orchid varieties means they're commonly found in garden centres and gift shops.

Orchids and symbolism

Orchids have long represented love, refinement, friendship, prosperity and beauty, which makes them an especially meaningful gift for a friend or loved one. They're a beautiful and long-lasting way to express your admiration or affection. Their importance isn't a new trend, however; some of the world's oldest civilisations prized orchids for their symbolic properties.

In ancient China, Confucius admired the resilience and beauty of orchids, growing "where others cannot". He compared it to "the life of the true gentleman... whose character shines no matter where he is or what he experiences".

In Japan, samurai warriors viewed them as an emblem of courage, while in Thailand orchids were frequently seen at weddings as they're associated with love, fertility and good fortune.

The Ancient Greeks are thought to be responsible for the name "orchid", adapted from the Greek "*órkhis*", which means "testicle". Not the word which immediately jumps to mind when you gaze upon your beautiful new plant? No, I didn't think so... However, there was logic to it – although our moth orchids have roots, many other orchid species have tubers, or underground stems, which resemble this particular part of the anatomy.

And so the name stuck, along with the Ancient Greeks' belief that eating an orchid could increase fertility or even determine the sex of a child. Men would eat large root tubers if they wanted a boy, while women would eat small root tubers in the hopes of conceiving a girl.

Don't fancy that particular taste test? How about a mixture of vanilla orchids and cocoa, an elixir created by the Aztecs which they believed would enhance strength and courage in battle.

> **Vanilla**
>
> This popular flavouring comes from a type of orchid – the only one to produce edible fruit. The vanilla bean is a seed pod of the vanilla orchid and is used to produce an organic compound called Vanillin, used in all kinds of baked goods, ice cream and perfume.

Nowadays, vanilla orchids (*Vanilla planifolia*) are still used to produce this popular flavouring, while other varieties can be used to create teas, or to garnish salads or desserts. They're even used in some high-end beauty products; scientists have been testing various species to identify their anti-ageing properties.

While "Orchidmania" may have died down, the international trade in orchids remains a million-pound business, and there's a booming black market for rare specimens fuelled by social media, threatening a number of species with extinction.

Over the years, orchids have made their way into popular culture too, with writers penning science fiction stories about killer plants; they've also featured in films and television, where they're often used to signify passion, sensuality or bring a sense of the exotic.

Despite their darker connotations on screen, orchids today convey a range of different sentiments according to their colour, either when given as gifts or used as decorations.

White flowers are seen as symbols of purity and innocence, given as a sign of respect, or to bring hope. Purple orchids can signify royalty and wealth, while yellow ones capture happiness and joy – a good flower to give someone to say "good luck" or "congratulations". Red – the colour of love and desire – says "I love you", whilst green orchids symbolise harmony and good fortune.

> **Blue orchids**
>
> Have you been given a blue orchid? These are special – but not in the way you might think. What you need to know about blue orchids is that they're extremely rare, and the likelihood of you having a real one is, unfortunately, extremely small.
>
> The blue orchids found on shop shelves are usually a white orchid which has been coloured by injecting blue dye into the stem. This turns the petals a vivid blue colour, which will last until the flowers fade. After that, the plant will re-bloom in its natural shade.

Origins and natural habitat

Over the years more than 30,000 species of orchids have been discovered, and breeders have developed hundreds of thousands of hybrids; it's one of the largest families of flowering plants and you can find them in almost every colour, shape and size, as well as in scented or unscented varieties.

> **Money doesn't grow on trees... but orchids do!**
>
> Most orchids grow in rainforests so they like typically tropical conditions: warm air, humidity and some shade, although some grow in cooler climates and they prefer lower temperatures. Many orchids, including Phalaenopsis, are epiphytic, which means they grow on trees, clinging onto the bark with aerial roots.
>
> This sounds like it would be bad for the tree, but it's not – the tree is merely used by the orchid for support and isn't damaged by the partnership. The orchid takes up water and minerals from the air and rain through thick, absorbent roots.

Varieties of orchid

This book will focus on Phalaenopsis, also known as the moth orchid. It's the most popular type of orchid, found in garden centres, supermarkets and florists, and is sometimes the first purchase which leads to a large and varied collection – a "gateway orchid", if you will!

Quite often when we buy them from shops and supermarkets, the label simply says "Orchid" or, if you're lucky, "Phalaenopsis Orchid". This gives the impression that they're all the same, but in fact, there are thousands of different hybrid varieties of moth orchid, although only a few hundred of these are mainstream cultivars which make it onto our shop shelves. They may have small differences in size, colour or flower shape, but they have all been specially bred by experts to develop plants which have all of the best characteristics of a moth orchid: long-flowering, colourful and tolerant of household conditions.

Whichever moth orchid you've got your hands on, you're sure to find it a rewarding plant to grow, because with some basic care and common sense, you'll enjoy flowers for weeks. And if you want to expand your collection, or step up to a more challenging specimen, there are lots of other Phalaenopsis varieties to seek out and to try growing.

This could be the beginning of a wonderful new orchid obsession!

Characteristics of the moth orchid

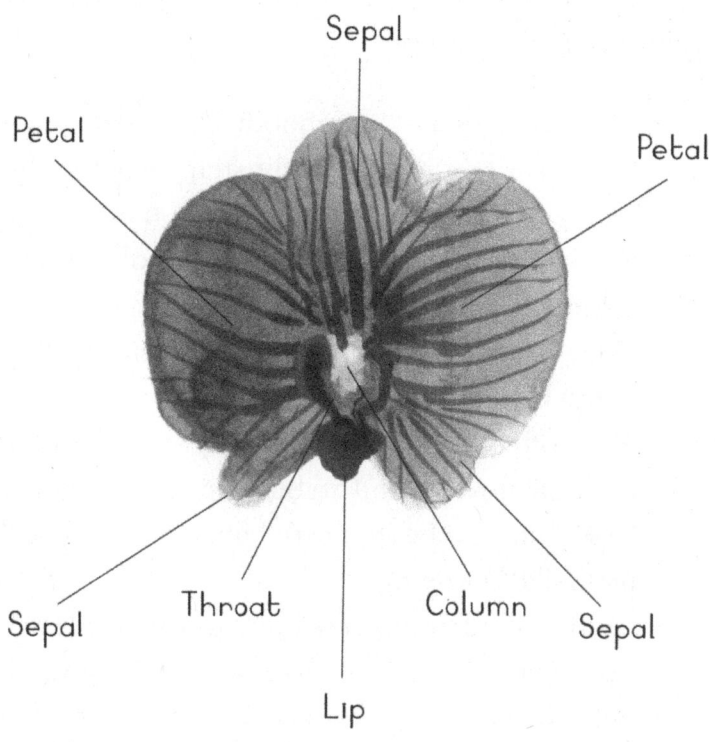

As we're focusing on Phalaenopsis — moth orchids — let's take a closer look at the common characteristics of this particular variety:

- Moth orchids are "monopodial", which means they have one main upright stem which holds all the leaves and flowering stems (rather than the leaves and stems coming out of the soil separately or in clumps).
- They have large, thick leaves grouped together at the base of the plant.
- They often have a long flowering stem, holding multiple flowers, and arching up and out of the base — this can be supported upright or may fall naturally to one side.
- Moth orchid flowers are quite distinctive. They have a flat face and generally round shape, all with the same basic pattern: each one has three sepals and three petals, and each side is a mirror image of the other. The sepals are the outer parts of the flower, protecting it when it's in bud; when it opens it reveals the two matching petals inside, and the lower lip. This is often a different shape or colour, and serves as a landing pad for pollinators, inviting them inside

the flower, where they'll find the pollen they're looking for!

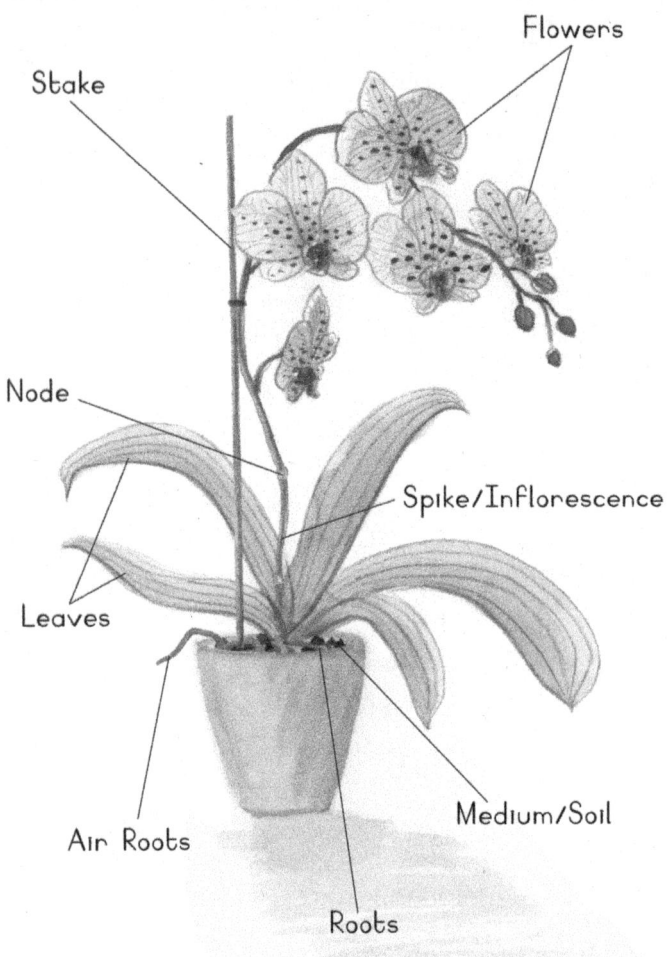

> ### A pet-friendly plant
>
> Unlike some houseplants, moth orchids are not toxic, and won't cause any harm to cats, dogs or other pets tempted to give it a nibble or brush past the flowers.
>
> However, you might want to keep the plant safe from especially inquisitive animals; a fall or sharp knock could dislodge a few flowers, and a light chewing might not do too much lasting harm but it won't make your moth orchid look great!

The basics: the essentials of plant care

There are a few general principles to bear in mind which will help you to keep your orchid alive, as well as any other plants in your possession.

Plants require three key things to stay alive and healthy: soil, water and light.

Soil

Almost every plant needs some kind of growing medium, something to put its roots into. This is most often commercial (or sometimes home-made) compost – a mix of organic material which has broken down over time and is rich in nitrogen and carbon, two nutrients which plants need to thrive. In the case of orchids, the growing element is usually bark – a chunky, open medium which provides enough moisture for the roots without leaving them waterlogged but is low in nutrients.

Water

This is essential for every plant, but the amount needed can vary widely, depending on species, size and season. Water is taken up by plants through their roots; this provides them with vital nutrients and is an important part of photosynthesis – the process by which plants make their own food inside their cells, which keeps them alive. Water also keeps a plant's roots, stems and leaves sturdy, as each cell holds the water, keeping it upright and growing in a healthy direction. Humidity is important for many plants, including orchids, which are used to growing in warm, damp environments and like to absorb moisture from the air around them.

Light

This is the other important element of photosynthesis – every plant's way of making the energy it needs to grow and to stay alive. Using light absorbed through its leaves, a plant turns water from the roots and carbon dioxide from the air into the food it needs to grow, releasing oxygen as a by-product. However, although it's essential, different amounts of light are needed by different plants; some are sun-lovers and need to bake in a hot spot for as long as possible – others need only a small amount of indirect light for a short time to meet their needs. As we've already learned, our orchids are somewhere in the middle: in keeping with their natural habitat up in the branches of trees, they enjoy bright light, but not full sun.

A few basic tools

To help you look after your plant and provide these essential elements, it helps to have a few basic tools, including:

- Watering can – for transporting water from the tap or water butt to the plant pot; it's also helpful if you want to mix some orchid feed in the correct amounts for fertilising your plant to help it thrive.
- Small trowel/fork – useful if you need to repot, for carefully levering out the orchid's roots or measuring out fresh bark/compost into the pot.
- Dish or decorative pot (something to set the plant into or onto) – setting your orchid's pot onto a dish of pebbles and water will help create a humid micro-climate for the plant; placing it into a decorative container can be more attractive than a plastic pot, although you may need to remove it periodically to inspect the roots and check if the plant needs watering.
- Cloth for dusting – moth orchid leaves are quite large and any dust will not only make your plant less attractive, but too much can hamper its

ability to absorb light, compromising the health of the plant; a soft cloth or even a pair of cotton gloves can be useful to keep the leaves free of dirt, and looking healthy and shiny.

Caring for moth orchids – or any other plant – is about providing the right soil, and the correct amounts of water and light. It's a simple formula which only needs to be tweaked slightly according to the time of year, or the age and stage of the plant.

Despite their looks, Phalaenopsis are not fussy or precious. Unlike some other houseplants, they won't keel over at the slightest draught or drop their leaves if moved an inch. They don't demand the consistent humidity of a high-end hotel spa, or wilt dramatically and fatally if you forget about them for a couple of days (I'm looking at you, ferns).

This is what makes orchids so desirable and rewarding as a houseplant. They're colourful, meaningful, non-toxic to pets, and they're long-lasting. A healthy plant will flower for two to three months, making them far more memorable and enjoyable than a bouquet of roses, which fades quickly by comparison.

Not only that, but they can bloom all year round and repeat flowering is typical, with the right care.

Plus, this tough little plant can withstand mistreatment, making a moth orchid an ideal gift for a green-fingered novice.

So perhaps instead of "How do I keep this alive?" we should be asking: "How might I kill this plant?"

2

HOW TO KILL THIS PLANT

Tips and tricks to ensure your orchid's certain death

There are a few sure-fire ways to mistreat an orchid to the extent that it will probably die. If you're really determined to commit vegicide you could try the following:

- *Dropping it from a third-floor window*
- *Spraying it with bleach*
- *Tying it to the back of your bike and dragging it down the street*
- *Allowing a dog to use it as a chew toy*
- *Creating a Rube-Goldberg machine which knocks down a series of marbles and dominos, ultimately releasing a bowling ball to crush it from a great height*

However, these are things which you'd have to deliberately go out of your way to do, in the manner of some kind of crazed plant murderer. There are other ways to kill it which are less obvious and more commonplace – the kind of mistakes which this book is going to help you to avoid!

Fortunately, it's actually quite difficult to kill an orchid by accident. Although they are a little different to other houseplants, they are fairly resilient and as long as you stick to a few simple instructions you should be able to keep it alive – potentially for quite a long time.

> ### KISS of Life
>
> Throughout this book you'll find a few basic care tips, helpfully highlighted in these "KISS of Life" boxes.
>
> KISS means Keep It Simple, Stupid. In other words, don't overcomplicate it – these simple, easy to follow instructions are the best way to ensure a good, long life for your orchid.

So what are those top tips? And, more importantly, what are the things you must definitely avoid in order to keep your orchid from becoming an ex-orchid?

Overwatering

Roots

Orchid roots are different from those of many other plants – look at how chunky and weirdly worm-like they are! Some might be a bit shrivelled and dry-looking too, but most of them, especially those inside the pot, should be quite turgid – you may even see the moisture shining on the surface of them.

However, despite their looks, these roots DO NOT want to be sitting in water all the time. Water-logging is a surefire way to end your moth orchid and is often the main cause of death by orchid murderers.

Avoid the following activities with your orchid:

- *Scuba diving*
- *Water polo*
- *Cold-water swimming*

If they're sitting in water, orchid roots will rot. Remember, these plants often have their roots hanging free in the wild, taking in moisture from the surrounding humid air. If the roots are sitting in water, they'll go mushy, brown and eventually die.

This is a mistake commonly made by new plant owners: the assumption that every plant needs lots of water. For orchids, every day is WAY too often and even watering once a week can be too much.

KISS of Life

green roots = no watering needed
silver-white or shrivelled roots = time to water
brown roots = hide the watering can

Luckily, moth orchids are equipped with an easy way to monitor when they are too dry, too wet or just right: take a look at their roots. It's one of the reasons why they usually come in clear plastic pots – so that we can clearly see the colour and condition of the base of the plant. Once the bark has dried out and the pot is lighter (both in colour and weight), check the roots: when they're green they don't need water; silver-white means they do. If the roots appear shrivelled, this means they've been under-watered. If they're soft and brown, the plant has been over-watered. To double check, it's also a good idea to get hands-on: assess the level of moisture by popping a finger into the pot and feeling whether the bark is dry or damp.

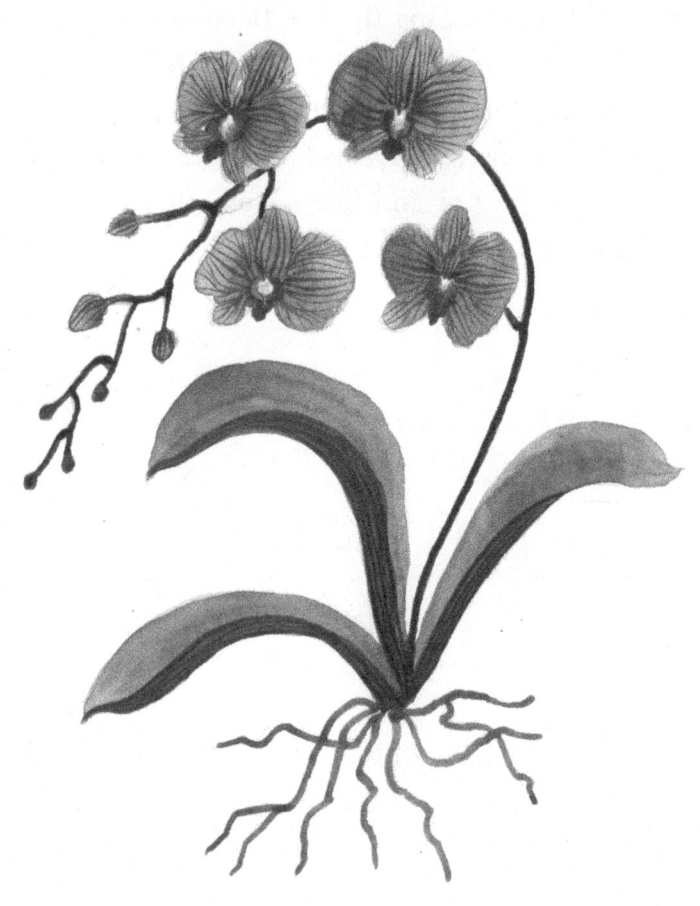

Leaves

Treat orchid leaves similarly: they like humid, damp air, but if they're wet for too long it can be harmful.

Misting is helpful for orchids as it can help replicate the humid conditions of their natural habitats – a plastic spray bottle filled with water will do the job, and they only need a light spray every couple of days. Problems can arise because the temperatures of our homes are not consistent, and if they're sprayed late in the day, followed by cool conditions during the night this means the excess moisture on the leaves won't evaporate. Try to avoid this, because it can provide the right conditions for bacteria to grow, leading to brown or black rot, infections which can take hold and damage your entire plant. The best thing to do is to mist lightly early in the day, which allows plenty of time for the leaves to dry off naturally.

> **KISS of Life**
>
> Keep the misting bottle close to the plant as a reminder, and give it a spray on Monday, Wednesday and Friday mornings (no need at the weekend – have a lie-in!).

Although some orchids need a period of rest, or a reduction in watering and feeding, this isn't necessary for Phalaenopsis, although they may benefit from being moved to a cooler room (16–18°C/60–65°F) during autumn.

So, we've established that too much water is bad for your plant. The best way to keep this plant alive is to ensure the compost and the leaves are kept moist, but never waterlogged.

> **KISS of Life**
>
> Water from the top, using a small watering can (or perhaps even directly from the tap – as long as you can keep the pot level in the sink).

Running water through the bark chips will flush out anything which is perhaps not harmful but is also not helpful – such as old fertiliser, salts or calcium (from hard water). Think of it as cleaning the roots and potting material, so they're in good condition to take up more of the good stuff: moisture, air and nutrients.

What kinds of water?

Tap water is fine, but if you're in a hard water area then soft water is better, and rainwater is best, as plants generally prefer its cleanliness and the natural balance of nutrients. Whatever you use, it's also beneficial for the water to be room temperature, as it hydrates roots better and faster compared to cold water, which can make the roots too cool, restricting the uptake of moisture.

Be aware: your moth orchid may have similar symptoms if it's been OVERwatered. If the roots have rotted away due to too much watering, the plant is still dehydrated, and it will have droopy, wrinkled leaves, because without healthy roots it has no method of taking up moisture.

If there are still some healthy roots present – or aerial roots growing up and out at the top of the pot – repot into fresh bark and reduce the frequency of watering. Use the roots as indicators – as described above.

If it appears there are no roots present at all, the plant is flatlining. There are a couple of techniques which may help: try suspending the orchid in a glass of water, ensuring the plant does not touch the water itself. The evaporation of the water will create humidity and should encourage new roots to begin growing in a few weeks.

🚨 ORCHID CPR 🚨

If your moth orchid is looking sad, with drooping, wrinkled leaves and those tell-tale shrivelled roots, it's dehydrated – but there is a way to revive it:

1. Fill a basin or sink with a couple of inches of tepid/room temperature water.
2. Place the plant in its plastic pot into the water for 20–30 minutes.
3. Allow the water to drain before placing the plastic pot back into its ornamental pot (if it has one).
4. Repeat after a week if the plant does not seem to have recovered.
5. If applicable, trim off any old flower stems to allow the plant to focus energy on its leaves.

Similar conditions can be created by placing a rootless orchid into a plastic bag with some damp moss or a wet paper towel in the opposite corner; whichever you use, make sure it isn't touching the orchid. Place the bag in a warm spot, but not in direct sunlight, and again roots should begin to develop in due course.

Once the new roots are a couple of inches long, you can repot the plant into fresh orchid compost and begin to water and care for it as normal. Be careful that you don't overwater, by keeping a close eye on the root colour and the surrounding bark, otherwise you could reverse all your good work and you'll be back to square one!

Too much light

Light is essential for plants, along with water and air. These three things combine to allow them to make the energy they need to grow and stay alive. However, there's such a thing as TOO MUCH or TOO LITTLE light. For example, it would be a bad idea to:

- *Keep your orchid in a cave*
- *Expose it to an overdose of gamma rays (you wouldn't like it when it's angry)*
- *Force your moth orchid into the spotlight (it could die of embarrassment)*

Orchids are tropical, and tropical plants love sunshine and hot temperatures, right? WRONG. In the wild, Phalaenopsis orchids grow on the bark of trees, on trunks and thick branches, so they can be quite high up, but are also shaded by other branches and leaves. This means they've got some of the lowest light requirements of their species, and they'll be quite happy on a shaded east-facing windowsill, or close to a west-facing window. They should be kept away from direct sunlight from spring to autumn. Strong direct sunlight can scorch leaves and cause the plant to overheat and wilt at any time of year, but especially in the warmer months. A north-facing windowsill is a good place to keep a moth orchid during summer.

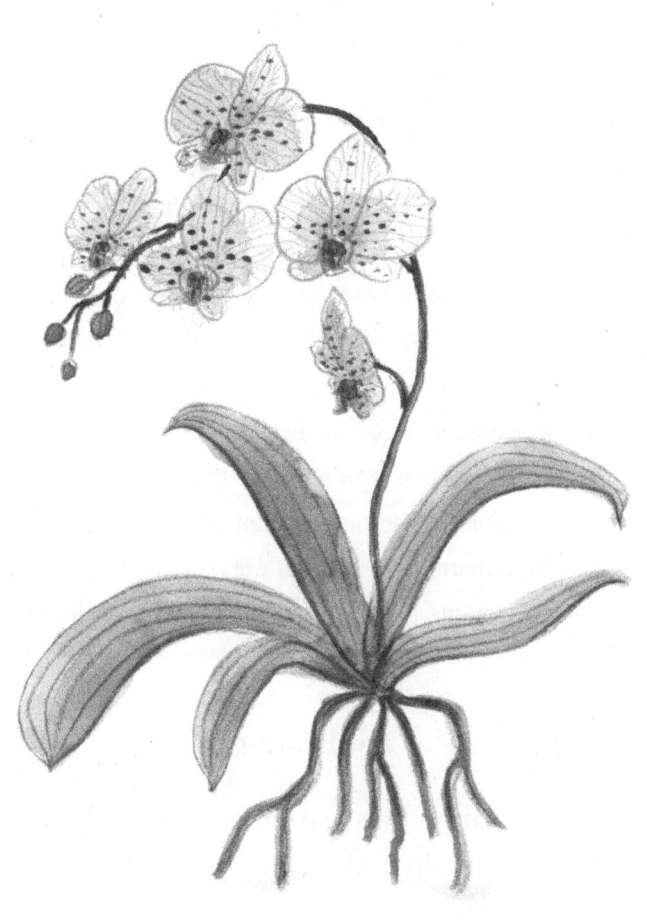

> ## 🚨 ORCHID CPR 🚨
>
> If you do manage to give your orchid a case of sunstroke, here's what to do:
>
> - Move it from its current position in bright sunlight – give it a more shaded spot
> - If the leaves have been damaged (black, white or brown marks are the telltale signs) then cut these back to healthy, green tissue
> - Make sure all other elements of its care are in place: regular watering, feeding, misting and – if possible – a humid atmosphere
> - Be patient – your orchid will recover, and should grow some healthy, fresh new leaves in time

Overheating

Although moth orchids can tolerate some dry conditions – both in their pots and the surrounding air – it's important not to overheat your orchid. Too much heat in any capacity will ensure its sudden and premature demise. Don't even think about:

- *Baking it in the oven like a Victoria sponge*

- *Styling it in a Stella McCartney faux-fur overcoat*
- *Setting a fire underneath it*

In the wild, your beautiful plant would be growing in and on trees, nestled into the bark, shaded by leaves and branches and cooled by consistent moisture; in our centrally-heated homes, if it's set beside a radiator, or given pride of place on your sunniest windowsill, an orchid can quickly develop the equivalent of heat-stroke – scorching and leaf collapse. Moth orchids are one of those houseplants which are most tolerant of the conditions inside our cosy homes, but they will struggle – and may give up the ghost – if they're exposed too long to high temperatures.

Conversely, temperatures which are too low can also affect moth orchids – it will stop the plant from flowering and you won't be able to enjoy its full beauty. Ideally, moth orchids want to be growing in daytime temperatures above 20°C/68°F and no lower than 16°C/61°F at night. Inside this range they should grow well and produce a reliable display of flowers.

Moth orchids really thrive if you can provide a warm environment that is also moist – as we learned previously, misting is a great option. Bonus points too if there is ventilation for some air movement, rather than dry and stuffy conditions. An easy way to do this is to set the pot onto a humidity tray. This is not as fancy as it sounds – all you need is a shallow plant-pot saucer or any container which will hold a couple of handfuls of gravel and an inch or two of water.

> **KISS of Life**
>
> Chances are, if you're too hot or too cold, then your orchid is too. Keep it in a room that would be a comfortable temperature for you to wear a T-shirt or light jumper.

Starvation

Food is important for orchids – their special compost is perfect for their roots and moisture absorption, but it doesn't provide much in the way of nutrition. You don't want to starve your moth orchid – but it's important not to overfeed it either. It's been rumoured that orchids don't like:

- *Fast food*
- *Dark chocolate digestive biscuits*
- *Low-fat yoghurt*

In its natural environment, high up in the tree canopies of rainforests, an orchid receives essential nutrients from rainwater and any surrounding organic matter, like decomposing leaves, animal droppings and other natural debris. This means it's used to receiving small, dilute amounts of food, and copes well with low-nutrient conditions.

In other words, if you only ever water your orchid and replace its potting bark every couple of years it should still survive for quite a long time. However, will it thrive? Probably not. Without added nutrients, plant growth will weaken. Older leaves may begin to yellow, and newer ones might develop slowly and be smaller than usual. It's likely that eventually flowering will slow down and could stop altogether.

There's little or no goodness in the kind of open bark compost which suits orchid roots best, so if you want big, healthy, glossy leaves and regular shows of flowers, then you must feed your plant to make sure it stays in tip-top condition.

Orchid fertiliser is widely available from garden centres and nurseries, or online. It comes in two different types, Grow or Bloom, both mixed with the correct balance of nutrients which will help your plant to achieve its full potential. Both varieties contain:

- Nitrogen – for stem growth and healthy leaves
- Phosphates – to stimulate good root growth
- Potash – encourages flowering

The Grow formula is designed to support an orchid when it's actively producing more leaves or roots, while the Bloom formula contains the right mix of nutrients to encourage it to flower.

> **KISS of Life**
>
> If you're likely to get mixed up about the best kind of food to use and when, then keep it simple: Bloom can be used all year round for Phalaenopsis orchids with no ill effects.

You can deliver the orchid food in a number of different ways:

- Diluting a concentrate into water and watering the plant with this mix. Fertiliser works best on damp roots, so if the compost is dry, make sure

you water it a little before watering the plant again with the addition of plant food.
- Spraying the leaves, roots and bark with a foliar feed (a foliar feed is one which is applied to the surface of the plant so nutrients can be taken up through the leaves and exposed roots). This has the added bonus of increasing humidity for the plant, which it will enjoy.
- Drip feeders – these are small containers of food which can be placed into the top of a pot, delivering fertiliser slowly and continuously over a period of time.

General purpose houseplant food can also be used, but it's important to note that this may need to be given at half-strength for orchids as it's considered too rich for sensitive orchid roots. This is why a specialist orchid feed is beneficial; it's formulated with the right balance of nutrients for these plants. The instructions will be clear and easy to follow, to help you avoid overdosing your orchid, which can lead to discoloured leaves and stunted growth.

In terms of frequency, feeding with every other watering is generally considered to be a good rule of thumb. And because overfeeding is more of an issue than underfeeding, it's fine to be relaxed about it; in other words, don't worry if you forget.

Eaten alive

Ensuring your orchid gets a good feed regularly is a good idea; equally important is making sure your orchid doesn't *become* a good feed.

There are a few pests which, if given free rein, will become silent assassins and could bring your orchid to an untimely end. You might want to look out for:

- *Ninjas*
- *Gremlins*
- *Borrowers*

Also watch out for Ancient Greeks gathering orchid tubers to eat for their alleged aphrodisiac, virility-boosting properties

It often takes a significant infestation to really kill off any plant, but pests of any kind can cause damage and impair the growth of leaves, stems and flowers.

Common pests of household plants, which could find a home in your orchid, include mealybugs, scale insects and aphids. All are quite small and may not be immediately obvious.

It's worth spending a little extra time every now and again closely inspecting your plant for any signs or symptoms of these tiny marauders. They're all there for the sap: the fluid which transports sugars and nutrients around plants. It's their primary food source, but it's also the plant equivalent of blood; these tiny insects attach themselves to leaves and stems to suck it out – plant vampires, if you will. Although they're viewed as killers, we can't really blame them. Like all living things, they're just trying to stay alive, get a good meal and repopulate their species.

Mealybugs – allowing mealybugs to run riot will result in distorted, yellowing leaves on your orchid, which may even fall off. If you spot fluffy white marks on the leaves, stem or the backs of flowers, this is a clear sign of mealybugs; they're the eggs and will soon hatch into sap-sucking adults. Check all the nooks and crannies of your plant, as these little critters like to hide away in crevices, such as the points where leaves and stem meet, or even in the bark inside the pot.

Scale insects – these don't even look like insects at first glance; they appear as small bumps or oval discs, usually on stems or leaves, but can also attach themselves to flowers. They're black or brown, and some are hard, others are soft.

Aphids – also known as greenfly, whitefly or blackfly. They're visible as little six-legged insects of various sizes, usually attached to the soft parts of the plant where they can easily extract sap. As they do this, they excrete a sticky substance called honeydew, which can become infected with sooty mould.

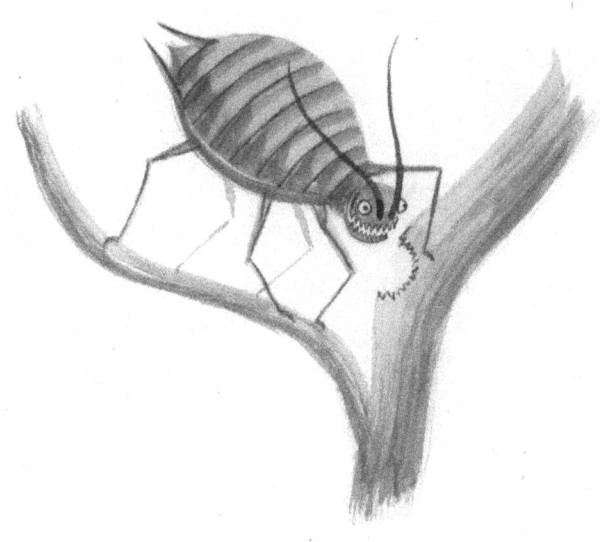

🚨 ORCHID CPR 🚨

If you spot pests on your orchid, it's not necessarily terminal. In most cases you can save the plant by removing the insects and employing good plant care practices to avoid future infestations.

When it comes to dealing with these kinds of pests you have a few options:

1. Get hands-on: you can rub off aphids, mealybugs or scale insects with your fingers or scrape them off with a thumbnail.
2. Spray them off: using a jet of water to physically remove them from the plant also works – just be careful not to scatter them further and spread them to other nearby plants!
3. Alcohol: not for you, it's for the pests. Applying alcohol or surgical spirit using a cotton bud (Q-tip) can help shift any beasties who are still clinging on.
4. Hired help: many pests can be controlled biologically using predatory ladybirds, lacewings or parasitic wasps, which you can buy and deploy as your own little insect army. This is a great way

> of letting nature do the work for you; however, it could be the horticultural equivalent of using sledgehammer to crack a nut if you only have a single plant and a small pest population. This approach is best if you have a larger collection or a more serious infestation.
>
> 5. Tolerate them: as stated above, pests don't have to mean a death sentence! Don't bin your plant at the first sign of a bug, or if they're proving difficult to eradicate. If it's a small, manageable population of pests, and as long as they're not impacting the plant's health, you can choose to live with them.

Also watch out for **fungus gnats**, also known as sciarid flies. These are the tiny, annoying black flies which live in the soil of houseplants and seem to take great pleasure in flying around your face and annoying you when you're trying to work/eat/watch TV. They aren't a deadly threat to orchids, but they multiply fast and are tricky to eradicate. It is possible to get rid of them, especially if you employ a few different methods to deter and control them, including:

- Yellow sticky traps – these can be hung near the plants or on a small cane inserted into the pot. They'll snag the adults and help to break the lifecycle. (Don't hang these outside, though, as you risk trapping beneficial pollinators such as butterflies, bees and hoverflies.)
- Nematodes – this is a biological control: predatory mites which kill the larvae, stopping the flies from reproducing. Packs of nematodes are diluted in water and applied to the soil of all your houseplants.
- Carnivorous companions – try growing an insect-eating carnivorous plant nearby. For example, a Drosera (or sundew) is good at trapping tiny flies in its sticky leaves. Plus, you can have the satisfaction of knowing that you're helping another plant to thrive by providing it with plenty of insect snacks!
- Allow your orchid to dry out between waterings and/or water only from the bottom by soaking the pot for a short time in a basin or sink. Fungus gnats lay their eggs in damp soil so they're much less likely to reproduce if the bark is mostly dry, or dry on the surface. In fact, if you do spot these little pests, it might be a sign

you're overwatering your orchid, so ease up on the watering can and see if that helps.

> **KISS of Life**
>
> Let's keep it simple: nip any pest problems in the bud by inspecting your plant weekly. Take a close look at stems, flowers and leaves, including the undersides, so that you can spot any suspicious visitors as soon as they appear and then take action.

Myth-busting: What won't *kill your orchid*

Finally, for would-be orchid murderers, a short guide to the things which are a waste of time if you want to put an end to your plant:

Planting it in an opaque container: Although the transparent pots are best for allowing plenty of light in and monitoring the appearance of the roots, planting your orchid in an opaque post won't kill it – it's just a bit more challenging to see the roots to find out if the plant needs to be watered.

Cutting down the flower stem: Even though an orchid without a flower looks a bit bare and sad, it's definitely not dead. As long as the leaves and roots are healthy, the plant will continue growing and should re-flower. This can happen quite quickly, or after a rest period; the key thing is to keep watering when it's dry, plus feeding with Grow formula, which will support the production of new leaves and help to maintain good health for existing leaves. In time another flowering stem will appear and bloom as before. In the meantime, learn to appreciate your plant's lovely green leaves!

Aerial roots: As in, the ones that grow out of the stem above the pot and float free of the compost. Don't worry about these – it's normal for some roots to grow up and out. You can mist these with water or orchid food to make sure they stay healthy and tuck them inside the container the next time you repot.

Repeat flowering: With the right care, the moth orchid can produce flower spikes holding multiple blooms as often as three times a year. This won't kill the plant, although it could exhaust it. If there are a lot of flowers and few leaves, or the flowers are in poor condition, consider cutting the flowering stem down as soon as it's finished blooming, and then feed your orchid with a Grow formulation food, to allow it to rest, recover and produce more leaves.

Ice cubes: Watering your orchid with ice cubes is a popular hack – placing three cubes at the top of the pot, so that the water is absorbed by the bark as they melt. The jury's out about whether this is an effective method or not, but what's certain is that it won't freeze your plant to death, despite its tropical origins.

3

HOW TO KEEP THIS PLANT ALIVE

Re-blooming, repotting and becoming a modern-day orchid collector

Easy wins for successful plant care

You've mastered the basics: how to keep your orchid alive. Yay! You're not a plant murderer! Well done, you.

You've hopefully discovered that your orchid is coping with its living conditions and responding well to regular water and plant food. Assuming you're not using it for target practice or misting it with acid, your plant is looking good and blooming happily in its new home.

By now you might be feeling proud of yourself, maybe even a bit smug. It really is quite easy to not kill your plant. However, "not killing" is still a fairly low bar. How about getting to grips with a few simple steps to help your plant *stay* alive for longer, so that it looks good, keeps blooming and perhaps even makes some plant babies?

After all, Phalaenopsis can live for about ten years or longer with the correct care and conditions, blooming twice a year for two to three months at a time. Some are recorded at fifteen or twenty years old. This plant could outlive your cat, your career or your current relationship!

So let's look at some further steps you might want to take, to ensure a long and healthy lifespan for your new plant pal.

Deadheading and re-flowering

What happens when your Phalaenopsis stops flowering? This period of rest for the plant is the classic point at which many orchid owners assume the plant is finished or that they've done something wrong and it's dying. It's flowered, now it's over, and it looks a bit boring and sad, so into the bin it goes. But all that's needed is patience.

After a few weeks in full bloom, the flowers will naturally wilt and fall from the stem. This is **totally normal** – you haven't killed your plant. As each bloom wilts simply pick them off the stem to keep the plant looking good; or you can wait for them to fall of their own accord, it's up to you.

Once the flowering spike is bare, you can encourage your moth orchid to re-bloom by shortening it down to a joint (also known as a node) further down the stem, leaving about 10–12 inches below. This will often prompt the growth of a new flowering stem from that joint; it will arch outwards from the main stem, and another cluster of flowers will develop and bloom on this new branch.

Alternatively, once the flowers have all fallen, you can cut the main flowering stem right down to its base. The plant will rest for a few weeks and might put out a couple of fresh leaves, then it should flower again. This process takes a bit longer but is likely to produce more flowers on the new stem growing from the base.

If the cut stem starts to go brown and looks unhealthy, trim it down to the base; or if nothing at all seems to be happening after a month or two, try placing the plant into a cooler environment for about six weeks before moving it back to a warm and bright position; sometimes a moth orchid needs a spell at a lower temperature to kickstart its flowering cycle again.

Keikei

Occasionally, another whole baby plant can grow from one of the stems of your Phalaenopsis! This is different from a second flowering stem, instead resembling a tiny version of the parent plant, complete with mini leaves and roots. It's called a "keikei", which is Hawaiian for "baby".

Now you have two options: you can cut this baby plant off and discard it – sounds harsh, but growing a second plant will eventually weaken the parent and delay its next flowering. So if more flowers are important to you, and you aren't bothered about owning a baby orchid, then don't feel bad about removing the keikei.

On the other hand, if you'd like to nurture the new offspring, wait for a short while, misting the keikei's roots with a foliar feed every day to help it along. It's ready to detach from the parent when it has two or three leaves and two or three roots of its own, at least three inches long. Snip off the keikei with about an inch of the attached parent stem and plant it into its own pot with orchid compost.

As it will only have small roots, overwatering is a risk, so monitor moisture levels carefully and keep spraying with orchid food.

Next: be patient! This little plant will take a few years to become full-sized and begin flowering, but it will be an exact copy of the parent. You've just *cloned a plant* – how cool is that?

> **Orchid ownership: a lesson in patience**
>
> It takes time for this plant to flower, whether you're encouraging it to re-bloom or nurturing a baby plant. But if you can continue to care for it and learn to appreciate its lush leafy looks as well as the exotic flowers, the reward – a long-lasting display of beautiful blooms – is well worth waiting for.

Repotting

Keeping your plant alive does come with some responsibilities; if you're really nailing those basics

– food, water, light and air – then sooner or later your orchid will be doing so well it will outgrow the container it came with. Then it's time to repot. In other words, upgrading your orchid's container so that its roots have room to stretch out, take in more good stuff from the soil and air, in turn producing more leaves and flowers.

Leaving it in the original pot **won't** kill it (at least not straight away!) – but it won't live up to its full potential either. In time, the roots will fill the pot, and the compost will break down, which means it won't be able to provide moisture and nutrients properly, and the plant will slowly lose condition.

That would take some time, however, so if you do spot the signs that your orchid needs repotting, don't panic! It's not a blue-light situation – but it's worth doing within a few weeks before the plant's health starts to suffer. Weaker plants are more susceptible to pests and disease.

So what are the signs to look out for?

- Lots of aerial roots will begin to escape out of the top of the pot – reaching out for more air and moisture which they're not getting from the compost. If all the roots inside and out look healthy, it's a sign your orchid is ready to be repotted. If the roots inside are brown then it's an overwatering problem – the compost is too wet and the roots are suffocating, so the plant is growing extra roots to go in search of air outside the pot.

- Very dark or black bark inside the pot – this happens when the compost breaks down and isn't draining well, so is holding more moisture than usual. Either way, you'll need to replace the compost to provide a healthier environment for the roots, with better drainage and more air.
- There are more roots than bark in the pot – if it's a tight squeeze, the roots won't be getting enough moisture or air. They need the bark compost to provide small pockets of air, and to absorb moisture which the roots can then take in.
- Sometimes the plant just *looks* too big for the pot. If there's a long, leafless section of stem rising up and out of the surface of the compost, that's a clear sign that it needs a deeper pot.

Orchids generally need repotting every two years or so, and it's best done in spring, preferably when the plant isn't in flower. Again, repotting when it's in flower **won't** kill it – it's just a bit more tricky, and you'll find you have to try and work around the flowering stem without damaging it and losing buds or flowers.

> ### KISS of Life
>
> Let's keep it simple: check the roots to find out if you need to repot. Too many inside or outside of the pot generally means you need a larger pot or to replace the bark.

A very short guide to repotting your moth orchid

In order to get started, you will need:

- **Fresh orchid compost** – use bark specifically made for orchids; this is readily available in garden centres and DIY shops with garden supplies. Don't be tempted to pop out to the local woods and collect some to make your own – commercial mixes contain the right blend of ingredients, usually bark and coir, which has been treated to remove any contaminants. It's clean, safe and will help your orchid stay healthy.
- **A larger pot** – not too large, though – you're aiming to go up one size larger than the old one (the exception is if you're trying to revive an orchid which has poor compost or damaged roots, in which case it can go back into the same pot with fresh compost). A clear pot is beneficial, but not essential – as we've already established, an opaque one will *not* kill your orchid.
- **A sterile knife or scissors** – for trimming long roots or cutting away any mushy or dead roots and old flower stems.

- **A spray** – water or food; aerial roots can be quite stiff, and spraying them with water or foliar feed can make them bend more easily, helping you to fit them into the new pot.
- **Watering can** – to give your plant a drink once it's settled into its new home.

1. If your orchid is quite dry, soak it for a few minutes. The plant will be easier to remove from the pot if the bark and roots are damp.
2. Squeezing the pot gently and holding onto the main stem, pull the plant from the pot, and carefully remove the old compost from around the roots.
3. Cut away any very long roots, trimming to a length of about 6 inches; also cut away any which are dead or rotting. A healthy orchid should have around 10–12 healthy-looking roots which are 2–3 inches long.
4. Spray any aerial roots to make them more pliable, then bend them downwards and set the stem and all the roots into the new pot.

5. Holding the plant in place in the centre of the pot, fill the spaces between the roots with the fresh compost. Give the pot a shake or a gentle tap to make sure the contents are settling and filling all the spaces – you don't want any large gaps in the compost or the roots may dry out.

6. Water the orchid after repotting and make sure the compost is draining correctly. Your orchid may not need watering quite as frequently now, as the pot is a little larger and may take longer to dry out. Bear in mind that you might have to adjust your watering schedule slightly.

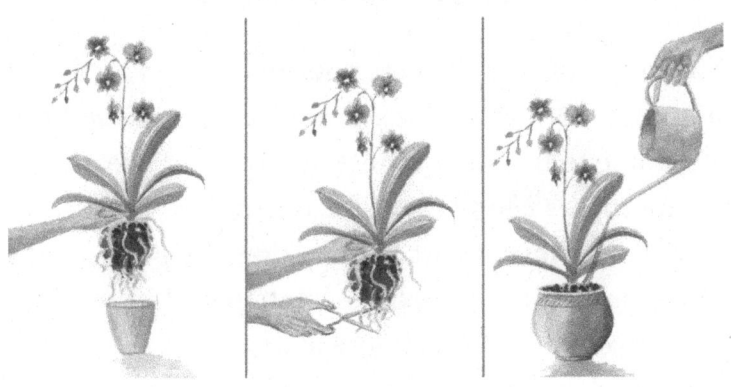

> **KISS of Life**
>
> When there are changes to the plant's location, pot or season of growth, keep checking the roots and the compost! Insert a finger into the bark and test to see whether it's dry or still damp.
>
> Remember the roots:
>
> - Green = no water needed yet
> - Silver-white = ready for watering
> - Grey and shrivelled = needs a thorough dunk for 20–30 minutes to restore moisture

Adding to your collection

So you've successfully kept your first orchid alive, and now you want to find it a friend? Of course you do! Welcome to the wonderful world of plant ownership, where there's *always* room for one more…

If you want to start a collection, or perhaps just test your new-found skills on another orchid, here are some tips to help you choose a new plant. All it takes is good observation skills and some common sense.

Let's start with the obvious – flowers. Your eye is immediately drawn to those plants in full flower, displaying all of their petals in beautiful colours. It's very tempting to pick up one of these showstoppers – and these are certainly great for instant impact – but they aren't always the best choice. It's impossible to tell how long the flowers have been blooming, and if they've been open and on display for some time, you may find they'll start to drop soon after you get the plant home.

Plants with a few closed buds at the tip are a better option, as these will gradually open and you (or the plant's recipient) will be able to enjoy their full beauty for much longer. Make sure the buds are firm, and avoid any plants with shrivelled or missing buds. Yellow or reddish buds are a bad sign too, indicating the plant has been stressed recently (too dry, too cold or too dark).

Next, take a look at the leaves: there should be three to four pairs and they should be firm and thick, with a healthy green colour. Avoid any plants which have leaves that are floppy or rubbery, or are a light green or yellow shade. Check for any signs of damage or pests (see the section on pests in Chapter 2 for more information on what you're looking for), and reject any which may have suspicious signs of infestation, including fluffy white spots or hard, bumpy growths on the surface.

Finally, inspect the roots. These should be silvery-grey with green tips, and nestling comfortably in some chunky orchid compost, in a transparent pot with good drainage (in other words, there should be holes in the bottom of the pot). If they look shrivelled and dry, or wet and brown, put that pot down and choose again. In fact, consider choosing a different seller, as this is a sign that the plants are not being well cared for!

It's not difficult to find a healthy, attractive plant in the shops or garden centres, but what about if you're shopping online? There are many great independent nurseries and orchid specialists who can supply plants by mail order, and some will also set up stalls at shows and markets for those who prefer to browse in person.

A reputable grower will ensure their orchids are sent out well protected for transportation, so they're not shocked by cold or damaged if manhandled. They also won't mind if a prospective buyer gets in touch to check delivery conditions beforehand or asks questions about their plants to ensure they're getting what they're looking for.

Specialist nurseries will often offer a greater range of colours and can provide specific details of each plant's name and variety. Some sellers will state if the plant is in full flower or not (these may even be reduced, as the flowering time will be limited); they're also a great place to source extra orchid kit, such as pots, plant food or bark.

The exception to the rule: Plant Rescue

Despite the advice given above about how to look for the best orchid, you'll inevitably find you get to a stage reached by every plant parent: ignoring all guidance completely and taking pity on the sad, rejected specimens found in the bargain bin.

These are where the most forlorn plants are to be found, with their wilting flowers, bare stems, dry compost and drooping leaves. However, these are often also the best bargains! Yes, they look a bit shabby and neglected, but they're often sold at a very reduced price and – as this book shows – it's actually very difficult to kill an orchid. So, if you're prepared to risk a small amount of money and provide some extra TLC, it's possible to revive some of the shabby specimens found on the supermarket shelves and transform them into healthy, flourishing plants.

In fact, it's a hugely rewarding process and makes you feel a bit like a superhero to bring home a neglected orchid, give it some care, and then watch it come back to life. It's one of the greatest joys to discover new, healthy roots or a fresh flower spike developing on a plant which looked like it was on its last legs just a few weeks ago.

Miniature Moths

Do you have a very small moth orchid? These are great little plants if you don't have a lot of space; however, some of these can be trickier to look after, so it's worth checking what you have, if you can.

Miniature hybrid varieties include:

Phalaenopsis equestris: A small and delicate moth orchid which is very easy to look after – it likes slightly drier conditions – and will produce lots of keikeis (baby plants).

Phalaenopsis violacea: Bears short spikes of fragrant starry flowers which are usually violet in colour, with white-green tips.

Phalaenopsis parishii: Another miniature variety, which needs a very humid environment and has fragrant flowers – but only when its growing conditions are exactly right!

4

OTHER COMMON ORCHIDS

More orchids you can give to others (or yourself), and how not to kill them

There are a few other orchids which you might buy or give as a gift, which, similarly to Phalaenopsis, are classed as "easy care", including Cambria, Cymbidium, Cattleya and Dendrobium orchids.

In this section, we'll look a little closer at these other varieties and how to keep them alive, should you find yourself expanding your orchid collection, whether by accident or design. Once you've got to grips with the basic principles of keeping a moth orchid alive, you should easily be able to transfer these skills to another variety, with a few tweaks, which we'll go into below...

Cambria

These are colourful, starry-flowered orchids, with slim, upright growth, which are hybrids of several different plants. They come in a wide range of bright colours, often with blotches and speckles, and a variety of shapes, including some with a large lower lip, sometimes ruffled, and others which have very long, spidery petals and are very unusual in appearance. If you want an orchid with distinctive looks, then check them out.

Most of these orchids are also scented – some more than others – and they all have long-lasting branching flower stems. Unlike moth orchids, with their single stem and thick roots, Cambria orchids have "pseudobulbs", from which grow multiple leaves and flower stems.

Temperature: Cambria prefer cooler conditions than Phalaenopsis. Their ideal temperature range is 10–24°C/50–75°F, so they're happiest in an unheated spare room or a spot far away from radiators and other heat sources.

Light: They do enjoy bright conditions but, as with Phalaenopsis, they should not be in direct sunlight during the spring and summer months as they could scorch.

Humidity: Cambria orchids need good humidity – spray them as you would do with a moth orchid.

Watering: These plants have much thinner roots, and, compared to moth orchids, are a bit more sensitive to too much or too little water. You should aim to keep them at a steady level of moisture; the exception is in the winter when it suits them to dry out a little between waterings.

Feeding: You can use the same Grow formula as you use for your Phalaenopsis on this orchid, all year round. If you want to level up, you can switch to Bloom during autumn and winter as this will improve the condition of the pseudobulbs and therefore its flowering ability.

Flowering: Cambria usually flower once a year in spring. The flowers are similarly long-lasting – about six to eight weeks – but won't re-flower like Phalaenopsis. Instead, the plant focuses its energy on new shoots which will grow into pseudobulbs to fuel the next set of flowers. Cut the stem down to the base once the blooms are all spent.

Cattleya

If you want something really flamboyant, then look to the Cattleya orchids. They have large, beautiful flowers which are usually lilac-pink with beautifully ruffled petals.

Temperature: They're from South America so they like it hot. 16–24°C/61–75°F is their ideal range, but don't be fooled – they don't like too much light as their leaves are more sensitive than other orchids and they can easily be scorched.

Light: Keep them on an east- or west-facing windowsill in the spring and summer, and make sure they have a bright spot in winter.

Humidity: Regular spraying is good for these orchids – mist the pseudobulbs, roots and bark, but avoid the leaves.

Watering: Regular watering is required in spring and summer to maintain a steady level of moisture but let Cattleya dry out between waterings in autumn and winter.

Feeding: Cattleya need a bit more nutrition than other orchids: they'll benefit from a Grow fertiliser with most waterings.

Flowering: These flowers will last three to four weeks, a little less than some other orchids, but just as impressive. After flowering it's best to remove the dead blooms and withered sheaths covering the pseudobulbs, as this helps prevent pests.

Cymbidium

A great subject for this book, as Cymbidium are said to be very difficult to kill! And they have a really impressive display of flowers, which can get bigger each year with the right care. They do have specific temperature requirements to make this happen.

Temperature: These are cool-growing orchids, preferring temperatures between 10–20°C/50–68°F, so they need to be kept in the coldest room of the house – an unheated conservatory is also a good place for them. They can be brought into a warmer position when they're in flower, so that you can enjoy the display. After they've finished flowering, Cymbidium orchids should be kept outside for the summer (as long as the nighttime temperatures are above 10°C/50°F). A period of low temperatures in late summer and autumn is what triggers the plant into re-flowering.

Light: Cymbidium orchids like good light all year round; an east- or west-facing position is usually fine, to provide a good spell of morning or evening light.

Humidity: These orchids will thrive in a humid environment, so mist regularly with a foliar feed as they also need lots of nutrients.

Feeding: You can use a Grow feed all year round with this plant, but it will also benefit from Bloom nutrients in autumn and winter.

Watering: Cymbidium prefer consistent moisture in the summer, but they should be allowed to dry out between waterings in autumn and winter. When the plant is outside in the summer keep a close eye on it – a dry and/or windy spell will dry it out quickly. It won't like being waterlogged either, so move it to a sheltered spot during a period of heavy rain.

Flowering: Large flowers in a wide range of colours – they'll grow bigger, with more flowering stems each year.

Dendrobium

These are also known as bamboo orchids because of the appearance of their stems, which look a bit like bamboo canes. To complicate matters slightly there are two types of Dendrobium, which each have different care requirements, and one of them looks a lot like the moth orchid, Phalaenopsis – in fact it's called *Dendrobium phalaenopsis*!

However, if you find one of these, you're unlikely to be confused for long, and you're in luck because it's also an easy plant to grow and care for.

Although the flowers are similar to moth orchid blooms, the growth habit of the Dendrobium version is quite different. Instead of a single, tall stem with leaves at the base, *Dendrobium phalaenopsis* produces a number of thin canes, with flowering spikes coming from the tips of these: usually one to four spikes with two to eight flowers on each.

Temperature: This plant enjoys similar growing conditions to its namesake: a warm room which doesn't dip below 15°C/59°F is ideal.

Light: A bright position is best, so sit this one on a south-facing windowsill in winter but make sure it's away from direct sunlight during spring and summer.

Humidity: These orchids love humidity, so spray them regularly and consider setting the pot on a humidity tray; a foliar feed sprayed on regularly is a good idea too.

Watering: As with other orchids, regular watering is a good idea, to keep the bark damp but not over-saturated. This plant slows down its growth a bit in the winter, so you can ease up on watering then.

Flowering: A bright environment encourages good flowering for this orchid, and it will often re-flower on old canes.

CONCLUSION

Enjoy your plant, keep it simple and avoid botanical brutality

So there you have it: in this book we've explored a lot of different ways you can kill your moth orchid. As you can see, it will take some effort to put an end to this plant. You will either have to think of a creative way to murder it, Agatha Christie-style, or commit to many, many weeks of neglect. Even then, administering CPR may still revive your orchid, bringing it back from the brink of death.

It's much easier, in fact, to simply try to keep it alive. Make the assumption from the outset that you're a good plant parent, and perfectly able to do so – because you are, and you can!

Even if you forget most of what you've read in this book, taking note of the **KISS of Life** top tips and following these few simple guidelines ensures you have all the knowledge you need to care for your orchid, ensuring it will grow, flower and stay healthy.

Remember the moth orchid's natural environment and how it would grow in the wild, taking nourishment from the air around it and the habitat of its host tree. In your home, *you* are the host tree, its life support: it needs you for water, food and the right living conditions.

Get to know your orchid – appreciate it regularly and get hands-on. You'll soon develop a sense of when it's happy and growing well, and when something's not quite right.

Clean, check and mist the leaves; insert a finger into the compost to check for moisture levels, and look closely at those tell-tale roots so you know when to water it. Give it a warm, bright spot in your home; remove potential pests and spent flowers to keep it looking good and staying healthy.

Try to avoid baking, freezing, drowning or allowing Ancient Greeks to devour your orchid.

It's as simple as that!

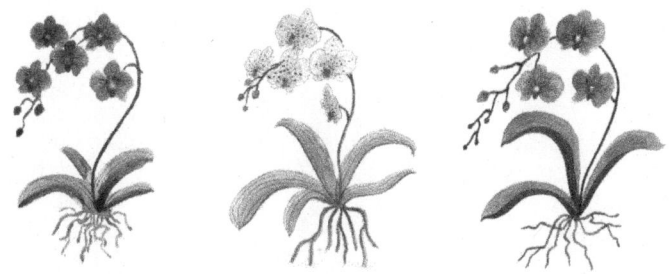